Beyond the World of Relativity
to the World of Invariance

BEYOND THE WORLD OF
RELATIVITY
TO THE WORLD OF
INVARIANCE

*A Journey of Discovery into
the Realm of Absolute Space and Time*

Thanh Giang Nguyen

Beyond the World of Relativity to the World of Invariance
A Journey of Discovery into the Realm of Absolute Space and Time

iUniverse books may be ordered through booksellers or by contacting:

iUniverse
1663 Liberty Drive
Bloomington, IN 47403
www.iuniverse.com
1-800-Authors (1-800-288-4677)

ISBN: 978-1-4917-8349-8 (sc)
ISBN: 978-1-4917-8348-1 (e)

Library of Congress Control Number: 2016905324

Print information available on the last page.

iUniverse rev. date: 03/28/2016

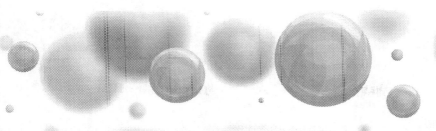

Contents

Introduction:
Discovering the World of Invariance

In the seventeenth century, Sir Isaac Newton and other natural philosophers developed classical mechanics, which became an accurate theory describing the motion of macroscopic objects under the action of forces. Then in the late nineteenth century, the Scottish mathematician and scientist James Clerk Maxwell proposed his theory of electromagnetism. He successfully predicted that an electromagnetic wave would propagate in free space with a constant speed.

The constancy of the speed of light was considered a point of conflict between electromagnetism and classical Newtonian mechanics, and some scientists devised new ideas to reconcile the discrepancy.

The most impressive of these was the theory of relativity published by the German-born American scientist Albert Einstein in the early twentieth century. After Einstein's theory had gained support from physicists, the equation $E = mc^2$ became a legend, and the theory of relativity became a pillar of modern physics.

The revolution in science that Einstein sparked was a revolution of *perspective*. The perspective of absolute space and time that had prevailed in Newtonian mechanics was displaced by the perspective of *relativistic space-time* in Einstein's theory.

While studying the theory of relativity, I realized it is not the only way in which the equation $E = mc^2$ can be derived. It can also be derived from fundamental physical concepts in Newtonian mechanics *within* the framework of absolute space and time, as outlined in this book.

I call this new idea that I have developed the theory of invariance.

In physics, every theory is a consistent conceptual picture describing the natural world. With the theory of relativity, the reality is a beautiful, multicolor, polygonal world. In contrast, the theory of invariance displays pure and delicate sceneries of nature.

So please join me on *the Ship of Invariance* to take a journey of discovery into a distinct world, a world of absolute space and time.

—Thanh Giang Nguyen

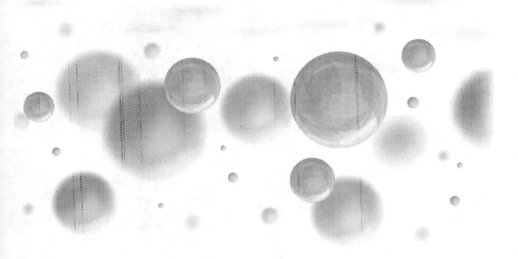

The Theory of Invariance
A Perspective of
Absolute Space and Time

1.
The Speed of Light

Speed is a fundamental concept in Newtonian mechanics. Since before the Galileo era, the speed of an object has been understood as the distance the object travels in a unit of time, and this conventional definition of speed has been applied to light since the eighteenth century. In the present, in science, the speed of light *in vacuum* is defined as *the distance light travels* in a unit of time in free space.

In the late nineteenth century, Maxwell's theory of electromagnetism showed that the speed of light is constant. This prediction was confirmed by the results of the Michelson-Morley experiments and de Sitter's binary star observations. The constancy of the speed of light to all observers became a great puzzle to scientists because the Newtonian mechanics could not allow the speed of an object to remain unchanged in all inertial frames of reference. In other words, the constancy of the speed of light was considered a point of conflict between electromagnetism and Newtonian mechanics.

In the early twentieth century, Albert Einstein introduced the special theory of relativity to reconcile the discrepancy. His revolutionary viewpoint of *relativistic space-time* played a crucial role in successfully establishing a mathematical link between the constancy of the speed of light and classical mechanics. However, special relativity contains an inconsistent point in itself. Speed, a fundamental concept in science, has been defined in two different ways. In the relativistic equation of Doppler effect, $f = f_o \sqrt{\dfrac{1 \pm v/c}{1 \mp v/c}}$

[1], *v* is *relative velocity* while the speed of light *c* is *conventional velocity*.

In contrast, the theory of invariance resolves the discrepancy by *modifying the definition* of the speed of light. In invariance, the speed of light is defined as *the rate of change of the distance* between the light and the object illuminated by the light source in a unit of time in free space. In other words, the velocity of light with respect to an observer defined in invariance is *the relative velocity* between them. With this definition of the speed of light, the theory of invariance becomes more consistent compared to special relativity because both velocities, *v* and *c*, are *relative velocity* in the invariant equation of Doppler effect. Besides, the results of the Michelson-Morley experiments and de Sitter's binary star observations are also appropriate with the constancy of the speed of light without conflicting with Newtonian mechanics. Thus, the nineteenth century's puzzle of discrepancy can be resolved without changing the structure of space and time to become relativistic. Interestingly, the classical perspective of absolute space and time allows us to derive the same form of the equivalence of matter and energy as that in relativity.

Let us examine why the speed of light is constant with respect to all observers from the perspective of absolute space and time.

Let us first consider a bicycle and a pedestrian that are in motion, heading toward and ultimately hitting each other. This experiment is repeated several times. It does not matter whether the bicycle or the pedestrian or both are moving. If the relative velocity between them is not changed, the pedestrian feels the same level of pain in every collision because the bicycle carries the same amounts of momentum and kinetic energy. If the relative velocity is greater, so is the pain the pedestrian feels because the bicycle now carries greater momentum and kinetic energy. Thus, the amounts of momentum and kinetic energy that an object carries relate directly with its relative velocity. A change in momentum and kinetic energy of an object will make a corresponding change in its relative velocity with respect to the observer.

Let us now consider an observer, a light source and photons emitted from this light source. Whether the observer is moving toward or away from the light source, that person still gets hit by the photons. When the observer is moving toward or away from the light source, he or she feels more or less effect of photons, respectively. It is the momentum and energy of photons that change when the observer is moving toward or away from the light source.

What happens to the photons themselves when their momentum and energy change? Photons are different from ordinary objects. Quantum mechanics states that momentum and energy of a photon are *proportional to its frequency* and, therefore, a change in their momentum and energy will make a corresponding change in their frequency *instead of* their relative velocity. This is the reason the relative velocity of light never changes no matter whether the observer is standing still or moving toward or away from the light source.

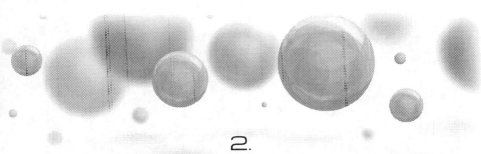

2.
Light Under the Effect of Gravity

Let us consider the following conceptual experiment:

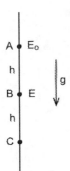

Figure 1: Light travels down from A to B
and C in a uniform gravitational field g.

A photon with energy E_o is emitted at point A in a uniform gravitational field g. Points B and C are positioned below the point A such that AB = BC = h. Call E the photon energy measured at point B, the energy E and the height h are variable amounts. Hence, the ratio (E/E_o) can be written as a function of the height, $\varphi(h)$, as follows:

$$\frac{E}{E_o} = \varphi(h) \text{, and } \frac{E_o}{E} = \varphi(-h).$$

Hence,

$$\varphi(h) \cdot \varphi(-h) = 1.$$

Call E_C the photon energy measured at point C, we have

$$E_C = E\varphi(h) = E_o\varphi(h)\cdot\varphi(h).$$

We also have

$$E_C = E_o\varphi(2h).$$

Hence,

$$\varphi(2h) = \varphi(h)\cdot\varphi(h).$$

Thus, $\varphi(h)$ is an exponential function:

$$\varphi(h) = e^{kh}$$

$$\frac{E}{E_o} = e^{kh}, \qquad (1)$$

$$\Delta E = E - E_o = E_o(e^{kh} - 1). \qquad (2)$$

Because the energy of a photon is proportional to its frequency, equation 1 can also be interpreted as a ratio of the frequencies of the photon:

$$\frac{f}{f_o} = e^{kh}. \qquad (3)$$

In real experiments, for values of gh significantly less than the square of the speed of light, the ratio of the frequencies is measured as

$$\frac{f}{f_o} \approx 1 + \frac{gh}{c^2}. \qquad (4)$$

Comparing equations 3 and 4, we obtain:

$$k = \frac{g}{c^2}.$$

Substituting $k = g/c^2$ into equations 1, 2 and 3, we obtain:

$$\frac{E}{E_o} = \exp\left(\frac{gh}{c^2}\right), \tag{5}$$

$$\Delta E = E_o\left[\exp\left(\frac{gh}{c^2}\right) - 1\right], \tag{6}$$

and

$$\frac{f}{f_o} = \exp\left(\frac{gh}{c^2}\right). \tag{7}$$

This equation describes the effect of gravity on light. However, in the universe, there is no uniform gravitational field, but only gravitational fields around astronomical bodies. For a spherical body of rest mass M_o, the frequency f_o of a light ray emitted at a distance R_o from the center of the body will be changed to frequency f as the light ray comes to a distance R from the center of the body:

$$f = f_o \exp\left[\frac{GM_o(R_o - R)}{R_o R c^2}\right], \tag{8}$$

where G is the gravitational constant.

Hence, if the light ray approaches infinity, then its initial frequency f_o will be reduced to frequency f_x:

$$f_\infty = f_o \exp\left(-\frac{GM_o}{R_o c^2}\right). \tag{9}$$

This equation implies that the frequency f_∞ approaches zero as R_o approaches zero. This consequence indicates that if a black hole exists, then it has no dimensions and no event horizon around it.

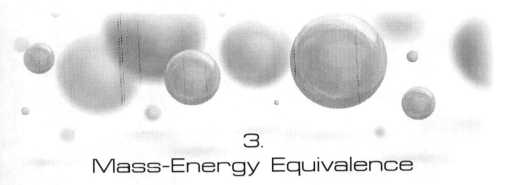

3.
Mass-Energy Equivalence

Let us consider an object m that is dropped freely from point A in a uniform gravitational field g. Points B and C are positioned below point A such that $AB = BC = h$.

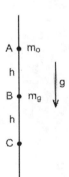

Figure 2: Object m falls freely from A to B and C in a uniform gravitational field g.

Call m_o the rest mass and m_g the gravitational mass of the object m. The gravitational mass m_g and the height h are variable amounts. Hence, we can write the ratio of the masses as a function of the height, $\varphi(h)$, as follows:

$$\frac{m_g}{m_o} = \varphi(h), \text{ and } \frac{m_o}{m_g} = \varphi(-h).$$

Hence,

$$\varphi(h) \cdot \varphi(-h) = 1.$$

Similar to the analysis used in section 2, we obtain:

$$\varphi(2h) = \varphi(h) \cdot \varphi(h).$$

Thus, $\varphi(h)$ is an exponential function:

$$\varphi(h) = e^{kh}.$$

Hence,

$$m_g = m_o \varphi(h) = m_o e^{kh}. \tag{10}$$

Call F the gravitational force that is exerting a pull on the object m in the gravitational field g,

$$F = m_g g.$$

From the definition of work, W, we obtain:

$$W = \int_0^h F dh = \int_0^h g m_g \, dh.$$

Substituting equation 10 into the equation above, we obtain:

$$W = \int_0^h g m_o e^{kh} dh$$

$$W = g m_o \frac{1}{k} \left(e^{kh} - 1 \right). \tag{11}$$

Now, let us consider the object m that "decays" into two "pieces" of light precisely when the object is dropped. One piece emits upward and the other piece emits downward. The upward-emitting piece immediately hits a mirror and reflects downward combining with the other piece. Call E_o the total amount of energy of both pieces of light measured at point A. Call E the total of energy of them measured at point B.

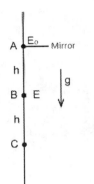

Figure 3: Object m "decays" into light precisely when dropped.

Comparing equations 6 and 11, let us pay attention to the portions $[\exp(gh/c^2) - 1]$ and $(e^{kh} - 1)$ in the equations, respectively. We obtain:

$$k = \frac{g}{c^2}.$$

Comparing equations 6 and 11, let us now pay attention to the portions (E_o) and (gm_o/k) in the equations, respectively. We obtain:

$$E_o = gm_o \frac{1}{k}.$$

Substituting $k = g/c^2$ into the equation above, we obtain:

$$E_o = m_o c^2. \tag{12}$$

This equation describes the mass-energy equivalence for an object at rest. Interestingly, this equivalence is the same as that in special relativity.

Now, substituting $k = g/c^2$ into equation 10, we obtain:

$$m_g = m_o \exp\left(\frac{gh}{c^2}\right). \tag{13}$$

Substituting equation 12 into equation 5, we obtain:

$$E = m_o c^2 \exp\left(\frac{gh}{c^2}\right). \tag{14}$$

Substituting equation 13 into equation 14, we now obtain:

$$E = m_g c^2. \tag{15}$$

Equations 13 and 14 describe the gravitational mass m_g and the mass-energy equivalence for the object m when it is falling through point B, respectively (see figure 2). Mass-energy equivalence for an object and its gravitational mass can also be expressed in its velocity. These relations will be derived in the following sections.

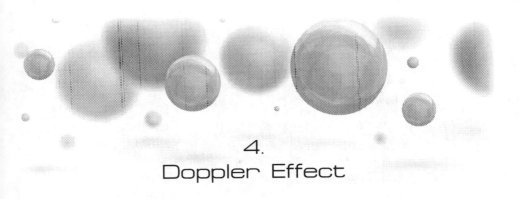

4.
Doppler Effect

Let us consider the following conceptual experiment:

Imagine a light source S and an observer O. The light source emits a flash of light toward the observer O. Call f_o the frequency of the flash received by the observer when at rest with respect to the light source. Call f the frequency of the flash received by the observer when moving at a velocity v toward the light source. The frequency f and the velocity v are variable amounts. Hence, we can write the ratio of the frequencies as a function of the velocity, $\varphi(v)$, as follows:

$$\frac{f}{f_o} = \varphi(v), \text{ and } \frac{f_o}{f} = \varphi(-v).$$

Hence,

$$\varphi(v) \cdot \varphi(-v) = 1.$$

Similar to the analysis used in sections 2, we obtain:

$$\varphi(2v) = \varphi(v) \cdot \varphi(v).$$

Thus, $\varphi(v)$ is an exponential function:

$$\varphi(v) = e^{kv}$$

$$\frac{f}{f_o} = e^{kv}. \tag{16}$$

In real experiments, for velocities significantly less than the speed of light, the ratio of the frequencies is measured as follows:

$$\frac{f}{f_o} \approx 1 + \frac{v}{c}. \tag{17}$$

Comparing equations 16 and 17, we obtain:

$$k = \frac{1}{c}.$$

Substituting $k = 1/c$ into equation 16, we obtain:

$$\frac{f}{f_o} = \exp\left(\frac{v}{c}\right). \tag{18}$$

This equation describes the Doppler effect for light. And because the energy of a light ray is proportional to its frequency, equation 18 can also be interpreted as

$$\frac{E}{E_o} = \exp\left(\frac{v}{c}\right). \tag{19}$$

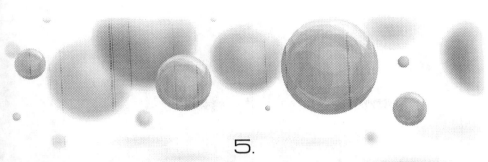

5.
Total Energy, Gravitational Mass, Kinetic Energy, Potential Energy and Linear Momentum

5.1. Total Energy

Let us consider the following conceptual experiment:

An object m is allowed to "decay" into two "pieces" of light when an observer is moving toward the object at a velocity v. Call m_o and E_o the rest mass and the rest energy of the object m, respectively. Applying equation 19 to the experiment, the observer receives a total amount of energy E of both pieces of light as follows:

$$E = \frac{1}{2} E_o \left[\exp\left(\frac{v}{c}\right) + \exp\left(-\frac{v}{c}\right) \right].$$

Using equation 12, we then obtain:

$$E = \frac{1}{2} m_o c^2 \left[\exp\left(\frac{v}{c}\right) + \exp\left(-\frac{v}{c}\right) \right]$$

$$E = m_o c^2 \cosh\frac{v}{c}. \qquad (20)$$

This equation describes the total energy of an object m moving at a velocity v.

5.2. Gravitational Mass

Comparing equations 15 and 20, we obtain:

$$m_g = m_o \cosh \frac{v}{c}. \tag{21}$$

In invariance, rest mass is different from gravitational mass. The rest mass m_o of an object m is measured by its resistance to being accelerated by a force when it is at rest. And the gravitational mass m_g is a measure of the strength of interaction of the object m with a gravitational field. Equation 21 above indicates that the gravitational mass of an object is dependent on its velocity. This consequence reflects a fact that when an object is in motion, it contains a greater amount of matter because it contains kinetic energy which is a form of matter.

5.3. Kinetic Energy

Kinetic energy is the difference between total energy and rest energy:

$$KE = E - E_o.$$

Substituting equations 15 and 12 into the equation above, we obtain:

$$KE = (m_g - m_o)c^2.$$

Using equation 21, we then obtain:

$$KE = m_o c^2 \left(\cosh \frac{v}{c} - 1 \right). \tag{22}$$

This equation describes the kinetic energy of an object m moving at a velocity v. It can be expanded as

$$KE = m_o c^2 \left[\frac{1}{2!} \left(\frac{v}{c} \right)^2 + \frac{1}{4!} \left(\frac{v}{c} \right)^4 + \frac{1}{6!} \left(\frac{v}{c} \right)^6 + \dots \right].$$

Hence, for velocities significantly less than the speed of light, this equation approaches the Newtonian mechanics kinetic energy equation.

$$KE \approx \frac{1}{2} m_o v^2 \text{ for } v \ll c.$$

5.4. Potential Energy

Let us return to section 3. Substituting $k = g/c^2$ into equation 11, we obtain:

$$W = m_o c^2 \left[\exp\left(\frac{gh}{c^2} \right) - 1 \right]. \tag{23}$$

By the law of conservation of energy, equation 23 also describes the potential energy PE of an object m at rest at height h in a gravitational field g as follows:

$$PE = m_o c^2 \left[\exp\left(\frac{gh}{c^2} \right) - 1 \right]. \tag{24}$$

Equation 24 is the invariant potential energy equation. It can be expanded as

$$PE = m_o c^2 \left[\frac{gh}{c^2} + \frac{1}{2!} \left(\frac{gh}{c^2} \right)^2 + \frac{1}{3!} \left(\frac{gh}{c^2} \right)^3 + \dots \right].$$

Hence, for values of gh significantly less than the square of the speed of light, this equation approaches the Newtonian mechanics potential energy equation.

$$PE \approx m_o gh \text{ for } gh << c^2 .$$

In general, the potential energy of an object m at rest at a distance R_1 with respect to a distance R_2 from a spherical object M is

$$PE = m_o c^2 \left\{ \exp\left[\frac{U(R_1) - U(R_2)}{c^2} \right] - 1 \right\} ,$$

where $U(R) = -\dfrac{GM_o}{R}$, and G is the gravitational constant.

5.5. Linear Momentum

From the experiment described in subsection 5.1, we can also determine the total amount of momentum p of both pieces of light with respect to the observer:

$$p = \frac{1}{2} \frac{E_o}{c} \left[\exp\left(\frac{v}{c} \right) - \exp\left(-\frac{v}{c} \right) \right]$$

$$p = \frac{E_o}{c} \sinh \frac{v}{c} .$$

Substituting equation 12 into the equation above, we obtain:

$$p = m_o c \sinh \frac{v}{c} . \tag{25}$$

In vector denotation, this equation is written as

$$\vec{p} = m_o \vec{v} \frac{\sinh(v/c)}{(v/c)} . \tag{26}$$

This equation describes the linear momentum of an object m moving at a velocity v. It can be expanded as

$$\vec{p} = m_o\vec{v}\left[1 + \frac{1}{3!}\left(\frac{v}{c}\right)^2 + \frac{1}{5!}\left(\frac{v}{c}\right)^4 + \ldots\right].$$

Hence, for velocities significantly less than the speed of light, this equation approaches the Newtonian mechanics momentum equation.

$$p \approx m_o v \text{ for } v \ll c.$$

Now from equations 20 and 25, we obtain:

$$E^2 - p^2c^2 = m_o^2 c^4\left[\cosh^2\left(\frac{v}{c}\right) - \sinh^2\left(\frac{v}{c}\right)\right]$$

$$E^2 = p^2c^2 + m_o^2c^4. \tag{27}$$

Even though the total energy and linear momentum described by equations 20 and 26 are different from those in special relativity, it is interesting that the total energy of a particle of rest mass m_o described by equation 27 is the same form as that in special relativity. However, they yield different values.

From equations 20, 21 and 25, we also recognize that the relations among the total energy, linear momentum and gravitational mass of a moving object m can be described as follows:

$$p = \frac{dE}{dv}$$

and

$$m_g = \frac{dp}{dv}. \tag{28}$$

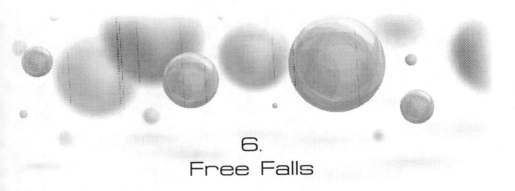

6.
Free Falls

6.1. Velocity

Applying the law of conservation of energy to the equations of potential energy (24) and kinetic energy (22), we obtain:

$$m_o c^2 \left[\exp\left(\frac{gh}{c^2}\right) - 1 \right] = m_o c^2 \left(\cosh\frac{v}{c} - 1 \right)$$

$$\cosh\frac{v}{c} = \exp\left(\frac{gh}{c^2}\right). \tag{29}$$

This equation describes the velocity v of an object dropped falling freely from a height h in a gravitational field g. It can be expanded as

$$1 + \frac{1}{2!}\left(\frac{v}{c}\right)^2 + \frac{1}{4!}\left(\frac{v}{c}\right)^4 + \ldots = 1 + \frac{gh}{c^2} + \frac{1}{2!}\left(\frac{gh}{c^2}\right)^2 + \ldots .$$

Thus, for values of gh significantly less than the square of the speed of light, this equation approaches the Newtonian mechanics equation:

$$v^2 \approx 2gh \text{ for } gh << c^2 .$$

6.2. Acceleration

Differentiating both sides of equation 29 with respect to time, we obtain:

$$\frac{1}{c}\sinh\frac{v}{c}\frac{dv}{dt} = \frac{g}{c^2}\exp\left(\frac{gh}{c^2}\right)\frac{dh}{dt}$$

$$a\sinh\frac{v}{c} = \frac{g}{c}\exp\left(\frac{gh}{c^2}\right)v .$$

Substituting equation 29 into the equation above, we obtain:

$$a\frac{\sinh(v/c)}{(v/c)} = g\cosh\frac{v}{c} \qquad (30)$$

$$a\frac{\tanh(v/c)}{(v/c)} = g . \qquad (31)$$

This equation describes the acceleration of an object falling at a velocity v in a gravitation field g. It can be expanded as

$$a\left[1 - \frac{1}{3}\left(\frac{v}{c}\right)^2 + \frac{2}{15}\left(\frac{v}{c}\right)^4 - ...\right] = g . \qquad (32)$$

Hence, for velocities significantly less than the speed of light, this equation approaches the Newtonian mechanics viewpoint about the relation between acceleration and gravitation.

$$a \approx g \text{ for } v \ll c \text{ and } a = g \text{ for } v = 0.$$

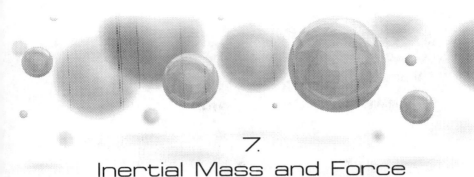

7.
Inertial Mass and Force

7.1. Inertial Mass

Let us consider an object m that is falling freely downward in a gravitational field g. Call m_i the inertial mass and call m_g the gravitational mass of the object. From the definition of inertial mass, $m_i = F/a$, and from the definition of gravitational mass, $m_g = F/g$, we obtain:

$$\frac{m_i}{m_g} = \frac{g}{a}. \tag{33}$$

Substituting equations 21 and 33 into equation 30, we obtain:

$$m_i = m_o \frac{\sinh(v/c)}{(v/c)}. \tag{34}$$

Hence, the equation of linear momentum (26) can also be written as

$$\vec{p} = m_i \vec{v}. \tag{35}$$

The inertial mass m_i of an object m is measured by its resistance to being accelerated by a force. Equation 34 indicates that the inertial mass of an object is dependent on its velocity. This consequence reflects a fact that when an object is in motion, it contains a greater amount of matter because it contains kinetic energy which is a form of matter. And equation 35 indicates that the linear momentum of an object can be defined as the product of its inertial mass and velocity. This definition is the same as that in Newtonian mechanics.

7.2. Force

In classical fundamental concepts, force is written as

$$F = m_i a$$

$$F = \frac{m_i m_g}{m_g} \frac{dv}{dt}.$$

Substituting equation 28 into the equation above, we obtain:

$$F = \frac{m_i}{m_g} \frac{dp}{dt}.$$

Hence, the law of conservation of linear momentum is not quite true, because, in invariance, inertial mass m_i is not identical to gravitational mass m_g. However, for objects moving at velocities significantly less than the speed of light, in a closed system, their inertial masses approach their gravitational masses and, therefore, the law becomes truer.

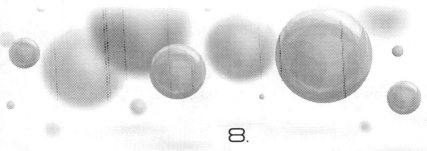

8.
Barycenter and Gravitational Force

Let us imagine two astronomical objects M_1 and M_2 that are orbiting around each other. Call R_1 and R_2 the distances from M_1 and M_2 to their barycenter, respectively. In Newtonian mechanics, the position of the barycenter and the gravitational force between M_1 and M_2 are described as follows:

$$M_1 R_1 = M_2 R_2$$

and

$$F = \frac{GM_1 M_2}{R^2}, \ [2]$$

where $R = R_1 + R_2$, and G is the gravitational constant.

Because the theory of invariance is based on classical concepts and perspective, invariant expressions of barycenter and gravitational force are very similar to those in Newtonian mechanics. However, gravitational masses are used in the expressions above instead of masses, because, by the definition of gravitational mass, gravitational mass is corresponding to gravitational force. Hence, the positions of the barycenter and the gravitational force are described by the following expressions:

$$M_{1g} R_1 = M_{2g} R_2$$

and

$$F = \frac{GM_{1g}M_{2g}}{R^2},$$

where

$$M_{1g} = M_{1o} \cosh\frac{v_1}{c}, \text{ and } M_{2g} = M_{2o} \cosh\frac{v_2}{c},$$

where v_1 and v_2 are the velocities of M_1 and M_2 in their orbits around the barycenter, respectively (see subsection 5.2, equation 21).

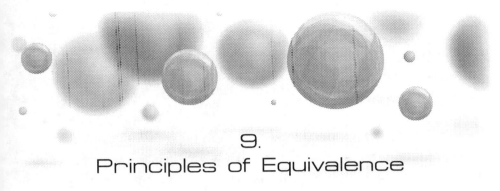

9.
Principles of Equivalence

This section examines current principles of equivalence from the perspective of the theory of invariance. The weak principle was established based on real experiments, and the strong principle was established based on Einstein's conceptual experiments.

9.1. The Weak Principle of Equivalence

The so-called weak principle of equivalence is also called the Galilean principle of equivalence. It states that all objects dropped freely in a gravitational field g will fall at the same acceleration a, and this acceleration a is exactly equivalent to the gravitation g numerically.

This weak principle is not quite true. Using equations 29 and 31, we can determine velocities and accelerations of falling objects dropped freely from a height in a gravitational field. Equation 32 shows that acceleration a is dependent on velocity v and *approximately equivalent* to gravitation g. In other words, the weak principle of equivalence is just an *approximate* principle from the viewpoint of the theory of invariance.

9.2. The Strong Principle of Equivalence

In 1907, Albert Einstein proposed a principle which is now called Einstein's principle of equivalence or the strong principle of equivalence. This principle can be stated in many different ways, and scientists say that all those ways are scientifically identical. Here, we will examine two ways of them and show that they are not identical. The first way is true and the second way is not quite true.

9.2.1. First way: For an observer inside a stationary laboratory sitting in a gravitational field *g*, the results of all local experiments are the same as those for an observer inside a laboratory moving upward at an acceleration *a* in free space, as long as *a* equals *g* numerically.

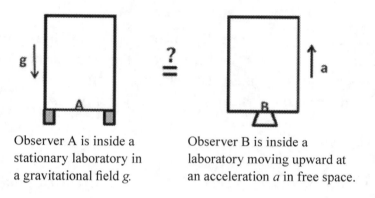

Observer A is inside a stationary laboratory in a gravitational field *g*.

Observer B is inside a laboratory moving upward at an acceleration *a* in free space.

Figure 4: The strong principle of equivalence stated in the first way. Is it a fact?

Installing two identical light sources into the ceilings of the laboratories, call *h* the height of the laboratories, and call f_o the frequency of the light pulses measured at the ceilings.

Let us first consider the left lab:

Using equation 7, we obtain the frequency of the light pulse received by the observer A on the floor as follows:

$$f_L = f_o \exp\left(\frac{gh}{c^2}\right).$$

Let us now consider the right lab:

Call $t_o = 0$ the time point that a light pulse is emitted from the light source on the ceiling, and call t the time point this light pulse is received by the observer B on the floor. Using the invariant definition of the speed of light, we obtain $t = (h/c)$.

The difference between the velocity of the floor at the time point t and the velocity of the light source at the time point t_o is $\Delta v = at = ah/c$. Using equation 18, we obtain the frequency of the light pulse on the floor as follows:

$$f_R = f_o \exp\left(\frac{\Delta v}{c}\right)$$

$$f_R = f_o \exp\left(\frac{ah}{c^2}\right).$$

Hence, if $g = a$, we obtain $f_L = f_R$. Thus, the strong principle of equivalence stated in subsection 9.2.1 is a natural fact.

9.2.2. Second way: For an observer inside a laboratory falling freely in a gravitational field g, the results of all local experiments are the same as those for an observer inside an unaccelerated laboratory in free space.

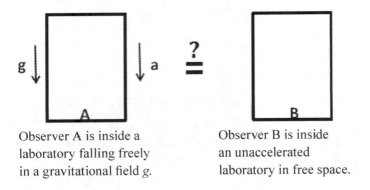

Observer A is inside a
laboratory falling freely
in a gravitational field g.

Observer B is inside
an unaccelerated
laboratory in free space.

Figure 5: The strong principle of equivalence
stated in the second way. Is it a fact?

Installing two identical light sources into the ceilings of the laboratories, call h the height of the laboratories, and call f_o the frequency of the light pulses measured at the ceilings.

This time, let us first consider the right lab:

This lab is unaccelerated and in free space, so the light pulse is not affected. The frequency of the light pulse received by the observer B on the floor is $f_R = f_o$.

Now, let us consider the left lab:

This lab is in gravitational field g and falling at an acceleration a, so the light pulse emitted will be affected by two factors, the gravitation g and the acceleration a. Similar to the analyses used in subsection 9.2.1, we obtain these factors as $\exp(gh/c^2)$ and $\exp(-ah/c^2)$. Hence, the frequency of the light pulse received by the observer A on the floor is

$$f_L = f_o \exp\left(\frac{gh}{c^2}\right) \exp\left(-\frac{ah}{c^2}\right)$$

$$f_L = f_o \exp\left(\frac{gh - ah}{c^2}\right).$$

Because acceleration a is approximately equivalent to gravitation g in a free fall (see equation 32, section 6), f_L is approximately equivalent to $f_o = f_R$. Hence, the strong principle of equivalence stated in subsection 9.2.2 is just an *approximate* principle from the viewpoint of the theory of invariance.

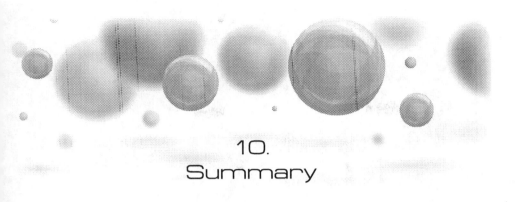

10.
Summary

The following are some interesting relations between masses, linear momentum and energy derived from this theory:

- The different masses of an object m are its rest mass m_o, which is defined as F/a when m is at rest,

$$\text{inertial mass } m_i = m_o \frac{\sinh(v/c)}{(v/c)}, \text{ and}$$

gravitational mass, which can be expressed as

$$m_g = \frac{dp}{dv} \text{ and } m_g = m_o \cosh\frac{v}{c}.$$

The relation between inertial mass and gravitational mass is

$$\frac{m_i}{m_g} = \frac{\tanh(v/c)}{(v/c)}.$$

The difference between inertial mass and gravitation mass indicates the conservation of linear momentum is not quite true.

- The linear momentum of an object m can be expressed as

$$p = m_i v, \quad p = m_o v \frac{\sinh(v/c)}{(v/c)} \quad \text{and} \quad p = \frac{dE}{dv}.$$

- The mass-energy equivalence is

$$E_o = m_o c^2 \text{ and } E = m_g c^2.$$

- The potential energy of an object m at rest at height h in a gravitational field g is

$$PE = m_o c^2 \left[\exp\left(\frac{gh}{c^2} \right) - 1 \right].$$

In general, the potential energy of an object m at rest at a distance R_1 with respect to a distance R_2 from a spherical object M is

$$PE = m_o c^2 \left\{ \exp\left[\frac{U(R_1) - U(R_2)}{c^2} \right] - 1 \right\},$$

where $U(R) = -\dfrac{GM_o}{R}$.

- The kinetic energy of an object m moving at a velocity v is

$$KE = m_o c^2 \left(\cosh \frac{v}{c} - 1 \right).$$

- The total energy of a particle m is

$$E^2 = p^2 c^2 + m_o^2 c^4.$$

- The Doppler effect for light is

$$f = f_o \exp\left(\pm \frac{v}{c}\right).$$

- The gravitational red/blue shift effect on light is

$$f = f_o \exp\left(\pm \frac{gh}{c^2}\right).$$

- A black hole, if it exists, is a point with no volume and no event horizon.

- The relation between acceleration and gravitation is

$$\frac{g}{a} = \frac{\tanh(v/c)}{(v/c)}.$$

- The position of the barycenter between two objects M_1 and M_2 is described as

$$M_{1g}R_1 = M_{2g}R_2.$$

- The gravitational force is

$$F = \frac{GM_{1g}M_{2g}}{R^2}.$$

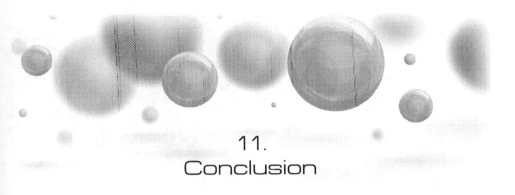

11.
Conclusion

We have studied the theory of invariance, which is based on the perspective of absolute space and time. In this theory, the definition of the speed of light is modified and, therefore, the speed can still be a constant with respect to all observers in free space without conflicting with Newtonian mechanics. Our analysis produces a series of invariant equations. For natural macroscopic phenomena, results calculated using invariant equations are approximately equivalent to results calculated using either Newtonian mechanics or special relativity. In addition, this theory allows us to reproduce two forms of the equivalence of matter and energy as seen in Einstein's special relativity ($E_o = m_o c^2$ and $E^2 = p^2 c^2 + m_o^2 c^4$). It also discovers the relations between acceleration and gravitation; and among rest mass, inertial mass and gravitational mass as well. These outcomes indicate that the theory can provide a proper view of the natural world, and the perspective of absolute space and time is an appropriate perspective in progressions of understanding the reality.

Throughout this journey, we have seen the natural sceneries that appear in the distinct world of invariance of absolute space and time are so delicate, pure and simple.

Thank you so much for reading with an open heart.

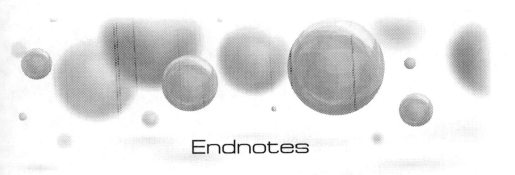

Endnotes

1. Edwin R. Jones and Richard L. Childers, 1994, *Contemporary College Physics*, (Reading, Massachusetts: Addison-Wesley), 714.

2. Arthur Beiser, 1991, *Physics*, (Reading, Massachusetts: Addison-Wesley), 210, 150.

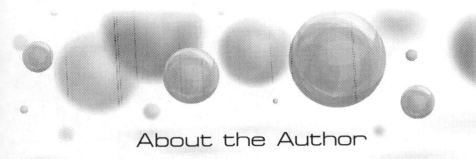

About the Author

Space and time are simply natural concepts.
They are non-materials.
They do not belong to the world of materials and,
therefore, they cannot be affected by materials.

—Thanh G. Nguyen

Thanh Nguyen disagrees with the mainstream perspective of relativistic space-time. Based on years of study and his own original work, he believes that time cannot be dilated and space cannot be curved by matter.

One could say that the author's distinct view on the invariability of space and time has been shaped by his many experiences with survival and death. Born in the Republic of Vietnam to a Bien-Hoa City police officer and a housewife, Nguyen witnessed the horrific consequences of the Vietnam War as a child.

In 1985, ten years after the fall of Saigon, Nguyen and thirty-eight others escaped communist Vietnam as refugees on a leaky, worn, five-by-twenty-foot wooden boat. This freedom came at a high price. For many stormy days and nights among dreadful waves and violent winds, these exhausted people wandered powerlessly and hopelessly into the vast Pacific Ocean and awaited what seemed to be inevitable and imminent death.

Nguyen still remembers vividly that when he and the others were drifting on the sea, the stormy nights were so dark that it was entirely black. The fierce gusty winds shrieked around, and the formidable angry waves towered over them. The suffering boat was wobbly escalating and then plunging nonstop in the relentless, roaring water. Attentively listening to miserable voices released from despondent people who were faintly groaning, crying and praying in despair, he sadly wondered from the bottom of his heart: What is the destiny of these people? Why is human life so frail? Why do we all have to die here, amid this immense ocean, buried under thousands of cold, huge waves?

After seven days and nights of struggling for survival, the refugees were finally rescued by a humanitarian captain and his team of sailors on a British commercial ship. They landed in Pusan City, Republic of Korea.

In 1987, Nguyen resettled in the United States. Once his new life became stable, he tried to find the answers to his questions in Buddhism, and he recognized and accepted the Buddhist beliefs of space and time are absolute.

In 1997, he enrolled at Worcester State College in Massachusetts, where he was introduced to the theory of relativity. He learned that most scientists support Albert Einstein's theory. Although Nguyen could accept the famous equation $E = mc^2$, he could not agree that space and time can be influenced because they are non-materials.

Continuing his studies independently, he developed his own way to derive the famous equation from his perspective of invariant space and time, which he has laid out scientifically in this book.

Nguyen's genuine views of human beings, society, the world and the universe have evolved in his heart and mind throughout his life. It is these views on life that have inspired his work.

Thanh Nguyen lives in Massachusetts with his wife and three children.